Whenever we buy, or sell, or pay, we're using money—every day!

1

Bike, kite, or treat—which would cost more if you saw these things inside a store?

Money comes in many a shape and size.
Which is which? Just use your eyes!

One dollar equals 100 pennies— and what about the other coins?

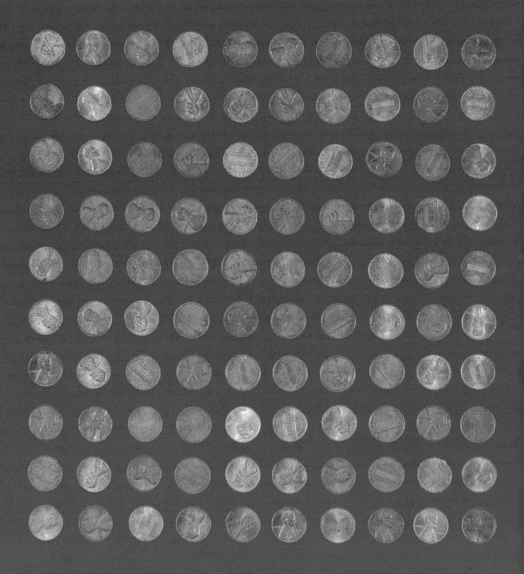

Let's count how many!

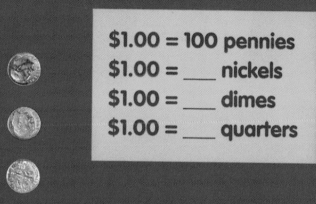

$1.00 = 100 pennies
$1.00 = ____ nickels
$1.00 = ____ dimes
$1.00 = ____ quarters

Read the prices on these things you can buy. How much is each? Give it a try!

How much will you pay for a pumpkin?

Let's count!
Four quarters and one dime
will make up the amount.

Before money is spent, it has to be made.
For work that we do, we can get paid!
We can rake, wash cars, or deliver the new
to earn the dollars and cents that we use.

The money you get
adds up over time.
So remember to save
each penny and dime!

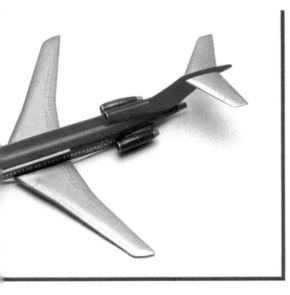

If you save up your money,
then you'll have more
to buy something special
you see at the store.

We use money in so many ways—
for things we need, or things for play.
But sometimes the best things we see
don't cost a thing—instead they're free!